THIS BOOK
BELONGS TO

Color Test Page

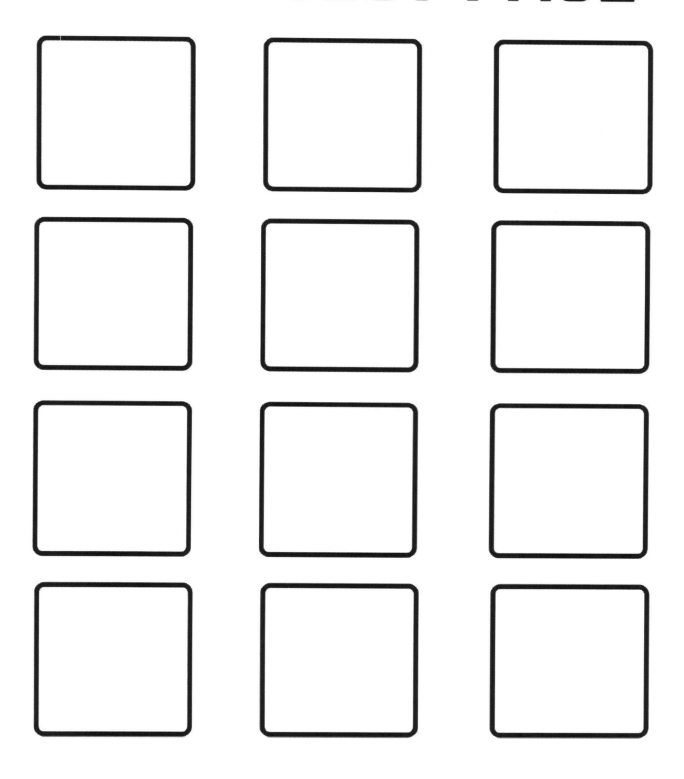

COLOR TEST PAGE

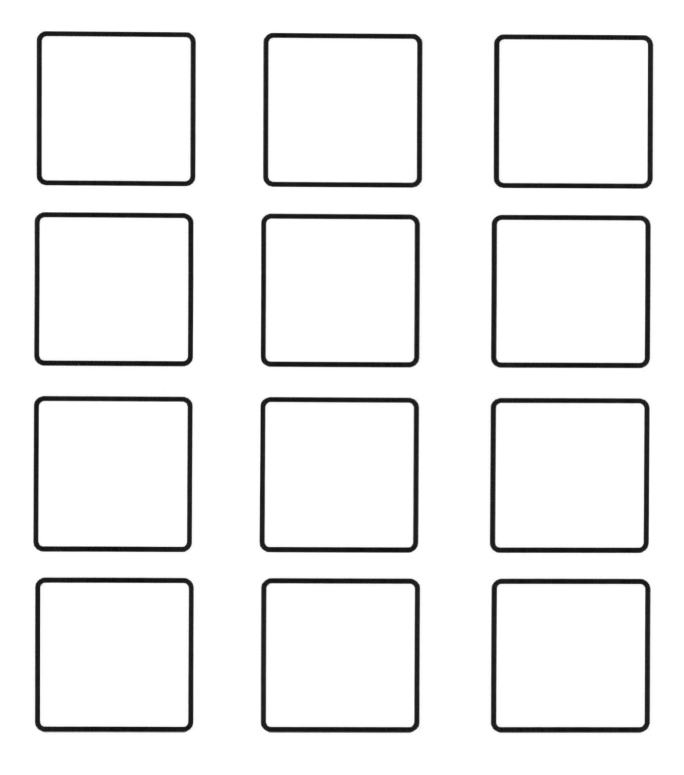

1-Red   2-Peach Cream   3-Congo Brown   4-Blue

1-Red   2-Peach Cream   3-Black   4-Geraldine

1-Pink   2-Peach Cream   3-Yellow   4-Purple   5-White

1-Pink   2-Peach Cream   3-Yellow   4-Purple   5-White

1-Turquoise   2-Peach Cream   3-Congo Brown   4-Lavender Purple   5-Baja White

1-Ship Cove   2-Peach Cream   3-Yellow   4-Tropical Blue   5-Red

1-Cabaret   2-Peach Cream   3-Congo Brown   4-Mandy   5-Coral Tree   6-Westar

1-Pink   2-Peach Cream   3-Congo Brown   4-Purple   5-Gray
6-White   7-Black

1-Red   2-Peach Cream   3-Congo Brown   4-Blue

1-Purple   2-Peach Cream   3-Brown   4-Green

1-Picton Blue   2-Dairy Cream   3-Driftwood   4-Pink

1-Purple   2-Dairy Cream   3-Pink   4-Black   5-Blue

1-Link Water   2-Dairy Cream   3-Pink   4-Brown

1-Scarlet   2-Milano Red   3-Black   4-Soft Pink

1-Pink   2-Green   3-Brown   4-Soft Pink

1-Purple   2-Green   3-Orange   4-Soft Pink

1-Purple   2-Soft Pink   3-Yellow   4-Red   5-White

1-Pink   2-Green   3-Brown   4-Soft Pink   5-Fountain Blue

1-Blue   2-Soft Pink   3-Brown   4-Red

1-Pink   2-Green   3-Brown   4-Soft Pink   5-White

1-Yellow   2-Soft Pink   3-Brown

1 violet 2 aqua 3 grey 4 blue grey
5 red 6 yellow 7 indigo 8 peach

Made in the USA
Middletown, DE
05 October 2023

40282905R00029